INDIAN SUPERSTITIONS AND TRADITIONS

AN ANCIENT SCIENCE

AKSHAY BAVDA

Copyright © Akshay Bavda
All Rights Reserved.

ISBN 978-1-63633-423-3

This book has been published with all efforts taken to make the material error-free after the consent of the author. However, the author and the publisher do not assume and hereby disclaim any liability to any party for any loss, damage, or disruption caused by errors or omissions, whether such errors or omissions result from negligence, accident, or any other cause.

While every effort has been made to avoid any mistake or omission, this publication is being sold on the condition and understanding that neither the author nor the publishers or printers would be liable in any manner to any person by reason of any mistake or omission in this publication or for any action taken or omitted to be taken or advice rendered or accepted on the basis of this work. For any defect in printing or binding the publishers will be liable only to replace the defective copy by another copy of this work then available.

TO MOM, DAD, VAISHALI AND RUCHI

Contents

Disclaimer — *vii*

Acknowledgements — *ix*

Preface — *xi*

1. Yogurt And Sugar — 1
2. Lemon And Chili — 3
3. Peepal Tree — 6
4. Black Cat — 9
5. The Bell Of The Temple — 12
6. Throwing Coins In Reservoirs — 14
7. The Day Of Cutting Hair And Washing The Head — 17
8. Bathing After A Funeral — 20
9. Broken Mirror Or Glass — 22
10. No Sweeping After Sunset — 24
11. Crushing A Snake's Head After Killing — 26
12. No Nail Cutting And Shaving After Sunset — 28
13. Forbidden To Plucking Flowers After Sunset — 30
14. Sleeping With An Iron Under A Pillow — 32
15. Keeping Head North Facing During Sleep — 34
16. Menstruation — 36
17. Eclipse — 39
18. Wearing Silk During Worship — 41
19. Vermilion — 43
20. Festival Of Deceased Relatives — 46
21. Cow Dung — 49

Contents

22. Namaskar	51
23. Navratri	53
24. Wearing A Bangles	56
25. Henna	58
26. Ear And Nose Piercing	60
27. Toe Ring	62
28. Tilak	64
29. Closing The Eyes Of The Corpse	66
30. Holy Basil	68
31. Diwali	71
32. Fasting	73
33. Gangajal	75
About The Author	77

Disclaimer

My purpose for writing this book is not to offend any individual, religion or society. To bring any country forward, the people of that country should adopt a scientific approach. This is just a small step towards cultivating a scientific approach.

Acknowledgements

A huge thank you to my MOM and DAD, I give all the credit for who I am today to my mom and dad. Those people have sacrificed a lot for me and made every effort to get me here.

My sister Vaishali inspired me to write this book. One day while we were talking on the phone, something happened that was associated with superstition. She told me that superstitious thing while talking on the phone. At the same time, in a quick reply from my mouth, I said, "There is nothing like that." As soon as she heard this answer, she started laughing and jokingly said to me, "Do one thing, brother, start writing a book explaining about superstition." So, these words were said to me in jest but I also got the idea that what Vaishali said is true. I must write a book on this. Then I told this idea to my wife Ruchi. Like me and Vaishali, She was also facing many problems due to superstitions so she smiled and said: " that's a very good idea, it will help society to move forward toward scientific approach for the superstitions". So, I started writing this book as soon as possible

When you have very few close friends you don't really have to name them: they know it without being mentioned. Thank you for being there. Sometimes that's all that matters. Last but not the least, love and gratitude to my family for standing by me and my decisions.

Preface

India is the only country in the world where people of many different religions live. Even centuries ago, India was at the forefront of science and technology. The credit goes to our sages, our sages who brought innumerable achievements to India. Science and technology are also the gift of the sages, who provided excellent examples of medical surgery that was not even possible in those days. At the same time, many achievements related to space also made India famous.

Our sages looked at the happenings around us from a scientific point of view and gave us the gift of very nice traditions. Even though sages know the scientific reasons for every tradition, I do not know why today traditions have turned into superstition. People may not fully understand the information or not properly pass it on from generation to generation. So today we are facing a lot of superstitions. Today, the scientific reasons behind almost all the traditions and superstitions have also disappeared.

Superstition is a subject that people are always running away from. There are many people who, despite knowing it, have accepted superstition just to avoid arguments with people who believe in superstitions. Our forefathers were very intelligent, so they converted the scientific fact into religion and put it into practice so that all would follow it unerringly. This beautiful tradition, which has been going on for years, did not take long to be spoiled. Years ago people were aware of the truth and adhered to scientific fact as a religious rule. But as the years went by, the scientific fact remained on one side and the religious rules on the other side and as a result this society got a unique gift of superstition.

PREFACE

We Indians are very joyful and enthusiastic creatures. It can be said that this birth as a Gujarati is obtained only when one should have done a lot of good deeds in the previous birth. But when it comes to superstition, we Indians also believe in superstitions.

Considering superstition, we can see three types of people. Almost everyone has heard a lot in their daily life that "Hey, don't do this and that, otherwise something bad will happen to you" This statement will introduce us to all three types of people.

The first type of people, who Utter sentences full of many different superstitions as written above and make the second and third types of people who live normally a little harder to live.

Then comes the second type of people, who hear the superstitious statement of the first type of people and accept that even though it is not true just because this type of person wants peace and doesn't like to argue. The third type includes those who rebel against the statement of superstition given by the first type of peoples.

The first type of people are great. Other types of people have to suffer a little bit but what can be done "दुनिया में अगर आये हैं तो जीना ही पड़ेगा" (Translation: If you come in the world, you will have to live). While the third type of people are the ones who are most often confronted with the most difficult situations, weapons like hatred, anger. why superstitious people support them? They make the offense of showing the courage to challenge superstition. Our heroes and heroines sometimes win in the battlefield of argument, sometimes they get defeated. So a line that is nicely set for them is "ज़िन्दगी हर कदम एक नयी जंग है"(Translation: life is a new battle every step you go ahead).

PREFACE

I was inspired to write this book because of the many occasions that have arisen around me. So, I thought that there will be many others like me, who can give this book as a gift to people who believe in superstitions. So,friends this is a gift I give you so you can also give this book as a gift to others it might help you to achieve little peace.If we observe precisely, a very long list of many superstitions can be prepared in our India. Not all of them are possible to cover in this book but I am presenting some of the most famous superstitions and traditions in this book with scientific facts as well as logic.

So, let's begin this auspicious journey by eating yogurt and sugar.

ONE

YOGURT AND SUGAR

""Superstition is the poison of the mind."
- Joseph Lewis "

One of the traditions or superstitions that has been going on for years is to consume a mixture of yogurt and sugar before going to do any good deed. This tradition has been going on since ancient times. A magical mixture of yogurt and sugar is given to a person before he goes to do any good deed. By consuming it, any good deed is completed very easily and peacefully. For example, even today, when a baby is going to take the first standard exam, his mother gives him a magical mixture of superstitious yogurt and sugar to make her baby score top rank in the first standard. The baby gets excited by the magical mixture she gave to drink and the baby happily goes to fight the exam.

Now in this era of stressful learning, through semester and grade marking, all the babies in the first standard are rewarded by "A" or "A" plus grades. The school does not hesitate to give good grades to all the students to prove the excellent teaching of their school. Babies think that good marks come from the magical mixture which was fed by mom before an examination, then a seed of this superstition is planted in their mind, and as time passes it turns into a tree of superstition.

So now it is necessary to know the scientific truth of this magical mixture. Yogurt is considered to be the best with medicinal properties in Ayurveda. Yogurt provides relief in many diseases. The good bacteria called Lactobacillus in it improves digestion and also gives coolness to the body. Sugar is an excellent source of glucose. Now the combination of these two provides energy to the body and calms the stomach and brain.

I am not saying that this combination is not beneficial but I am saying that the point of view is different. This is an excellent combination from a scientific point of view. If a person's stomach is calm then that person is calm and if a person's stomach is upset then the person is upset. In this way, yogurt calms the stomach so that a person can concentrate better, the glucose in sugar provides energy to the body so that the person can do their work with peace of mind and energy.

So when we are going to do something good, we must have eaten this magical mixture too. Yes, we may have eaten this mixture with a smiling face, but we have never thought about the fact that there is a scientific reason behind this superstition.

TWO
LEMON AND CHILI

""Superstition is the death of the thinking mind."
- Dr. T.P.Chia"

If we ever had to go to the market early in the morning, many of us would have seen a person selling a composition of lemons and chilies pierced with thread at the front and back of his bicycle. All the shopkeepers give him some money and buy this lemon chili and tie it on the door of their shop. If you asked about the reason, you will hear a nice statement that has been going on for years, "By hanging this lemon chili will save my family and business from evil eye and my business will run well."

If this has been going on for years, there must be something in it. I don't want to criticize anyone. But I have only one question to ask such peoples. Does this composition save you from a recession? Unfortunately, we do not have enough big lemons and chilies in our India to eradicate the effects of recession.

I'm sorry, the above statement has become a little too harsh. But after knowing the scientific truth, even you will be surprised that there is such a reason. Our ancestors were very intelligent and far ahead in the field of science. So they used lemon and chili as insecticides. Yes, you read that right.

Our ancestors used to make a composition by piercing lemon and chili in a cotton cord. The design was based on the scientific principle of capillary attraction (the attraction of a fluid in the capillary). So the capsaicin in the chili and lemonin in lemon spreads a special kind of strong odor in the atmosphere through the cotton thread. The odor is negligible to humans, but is very strong and deadly to small organisms such as flies, mosquitoes, and small insects. So that the insects run away from it and this composition acts as an excellent insecticide.

Years ago people used to hang lemon chilies in their bullock carts as well as at home. There is also a scientific reason behind this, in the past, due to the lack of electricity facilities, darkness prevailed after sunset. So if an insect bites, it can be identified with the help of lemon and chili whether the bitten organism is toxic or non-toxic.

After reading this, you may be wondering how to know whether an organism is toxic or non-toxic with the help of lemon chili?

When a venomous creature like a snake or a scorpion bites a human, its venom affects the human nerves and the person's taste sensing stops working. So that by eating lemon or chili on an immediate basis, one can know whether the creature that was bitten was poisonous or non-toxic.

So, in earlier times lemons and chilies were also hanged in bullock carts. Our ancestors may have been aware of this scientific fact but as the years passed this pesticide formulation took the form of superstition and even today in the 21st century we still believe in this superstition.

THREE
PEEPAL TREE

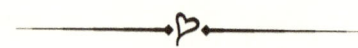

"Superstition are habits rather than a beliefs."
- **Marlene Dietrich**

The Peepal tree is given a different significance in Hindu culture than all other trees. Peepal is considered a very sacred tree and is also worshiped. Many of us may have gone with our mother to give water to the Peepal tree. You may have also heard many new stories about the Peepal tree.

In India, it has been believed for centuries that the Peepal tree is the symbol of "Pitru". Some people even believe in the belief that ghosts as well as spirits reside on the Peepal tree. During "Pitrumas" people worshipped Peepal trees. Over this time the Peepal tree is watered and it is believed that doing so brings peace to the soul of the Pitru.

Scientists have proven through their research that the Peepal tree produces much more oxygen than any other tree. Peepal trees can uptake carbon dioxide gas during the night as well because of their ability to perform a type of photosynthesis called Crassulacean Acid Metabolism. This type of process produces more oxygen.

The trick of our ancestors was also very beneficial to make the Peepal tree a religiously sacred tree. By associating Peepal trees with religion, people will stop harming the Peepal tree. So that there will be a very good amount of oxygen in the vicinity of the Peepal tree. One of the properties of peepal trees is that it purifies the air and destroys harmful bacteria in the air. During worshipping people do Circumnavigation around Peepal tree the reason behind this is to get pure oxygen.

Many people believe that Peepal trees are residents of ghosts. When it comes to ghosts, the Peepal tree produces a lot of oxygen. So to protect the Peepal from those who do not consider it sacred, our ancestors spread rumors that the Peepal has ghosts on it. Thus, if a person sits under any tree at night, he has to face chest pain and difficulty in breathing.

But the reason behind it is not a ghost or a spirit, the responsible reason for this is the tree produces carbon dioxide gas during respiration. Carbon dioxide gas is heavier than air so that it accumulates at the bottom. So if a person sits under a tree at night or falls asleep, he has to face problems like chest pain, not being able to breathe. So that people feel as if someone is sitting on their chest and also feel as if someone is trying to kill them by choking.

The seeds of superstition are sown by the Horrifying experience of such heroic people and by the spread of such news. No one even goes there to confirm the truth since one has experienced it. So the seed of this thing becomes a giant tree of superstition. Day by day people are watering this tree in the form of adding their rumors and as a result this has taken a very big form and the shadow of this dark superstition has been passed down to generations.

But it is our responsibility to remove this superstitious shadow and replace it with luminance of science. Our next-generation should be exposed to the truth from an early age to cultivate a scientific approach. Our forefathers had a very far-sighted thinking, so today we can see trees like peepal and banyan tree. We also have to make the next generation aware so that nature is protected and pollution is reduced.

Trees are an excellent source of pollution control. That is why to give a better life to our future generation, we should plant trees from today and reduce the use of plastic, the biggest enemy of nature.

Glossaries

- **Pitru**– Deceased loved ones/ Relatives
- **Pitrumas**– Worshiping month for deceased relatives

FOUR

BLACK CAT

""What the mind Doesn't understand, it worships or fears "
- Alice Walker "

The black cat can be called the most infamous animal in the world. It has been believed for years that when a person goes to work or just goes on the road and a black cat crosses his path, something bad happens to that person. As well as the person who is out to work does not get the job done.

There is no doubt that almost everyone has heard of this superstition. Whenever we go out and a black or any cat passes by our path, people stand there. It is believed that if a cat crosses the path, only after another person who is not present there passes by after that the person whose path the cat has crossed can proceed. If this is not done then there is no omen or something bad will happen. But the truth is that no one is harmed or any omen does not happen.

This superstition of crossing off the path by the cat is very prevalent. Even today, in the 21st century, almost everyone believes in this superstition. Unfortunately, no one has even tried or thought to know the actual reason behind this superstition.

In earlier times our ancestors had very little vehicle facility. They mostly used equipment like bullock carts or horse carts for transportation. Because of farming, some people had to go to the farm at night to water the plants. During this time there was lake of electricity so ,there was little light on the roads so the only way to get through the dark was to use a small light or lantern.

During that time when people are returning from the farm or another village in their ox cart or horse cart, suddenly the cattle get frightened. This is because of the black cat because the cat's eyes start to glow at night time and because of the black color only the sparkling eyes can be seen. So that animals like oxen or horses get scared of this scene, they get the impression that a beast or prey is coming. So those people didn't move forward and just stood there and sometimes the cattle were so scared that they got out of control.

Seeing this scene as well as seeing the bulls or horses scared, that person looks at the road and sees why this animal is scared. Observations show that a black cat is passing by. After this incident, whenever people saw a black cat from a distance they parked their cart there. So as not to frighten their cattle, as well as to avoid causing an accident due to loss of control over the cattle, they found the solution to stand and wait until another vehicle passes smoothly and makes sure that the road is clear.

Thus, due to such cases years ago, it was implemented that if a black cat is crossing the road, the person should be stopped. Don't go through the place until the cat is gone and another person passes by. As the years went by and people followed this rule, the real reason behind the rule gradually disappeared, every cat took the place of the black cat and we got a superstitious gift.

Today People believe in this superstition. People drive cars with the latest technology, bullock carts have stopped appearing, but today, whenever a cat crosses a person's path, he superstitiously stands as if he is driving a bullock cart and blindly follows the rules made by his ancestors.

There is no need to be scared or frightened if the cat crosses the road. There is a need to cultivate a scientific approach. It may be possible that the animal is in a hurry more than you and has to reach somewhere in time.

FIVE

The bell of the temple

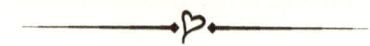

> *"The nation's culture resides in the hearts and in the soul of it's people"*
> *-Mahatma Gandhi*

Temple is a place where peace of mind is achieved, as well as positive energy, is also achieved. We are very well versed in the construction of the temple. Let us focus on a straight point rather than anything else about the construction of the temple, we must have seen that bells are installed in every temple. Every devotee in the temple worships their God and rings the bell together.

It is believed that chanting in front of the idol of the deity in the temple soon completes the desires. Everyone in the temple puts their desires in front of God and rings the bell with faith. When small children are attracted by the sound of clang. This is one of our traditions that has been going on for years.

There is a great scientific reason behind this tradition. The bells used in the temple are excellent examples of our ancient metal art. The bell is made very nicely by mixing different metals like zinc, copper, cadmium, bronze, etc. It is also shaped in such a way that it can produce excellent clang.

This type of bell, which has been used scientifically since ancient times, produces a clang for about seven seconds. It has been proven that this clang produced from the bells of the temple immediately balances the left and right brains of human beings so that the person realizes peace. This clang, which is generated for seven seconds, provides energy to seven chakras and also activates the seven chakras in human beings.

Apart from the temple, the bells that are rung during worship at home also work on the same scientific principle. Due to its small size, it produces a very loud clang, which has the advantage of destroying the harmful bacteria caused by the vibration and noise produced by its clang.

The tradition of ringing bells in the temple has been going on for years and still we should continue to do so for years. But yes, we should know the proper scientific reason behind it and also tell it to our coming generation. So that people also accept this tradition and this tradition does not turn into superstition.

SIX

THROWING COINS IN RESERVOIRS

> *"Fear is the main source of superstition and one of the main source of cruelty. To Conquer fear is the beginning of wisdom."*
>
> *- Bertrand Russell*

We must have heard this sentence, "If we will put a coin in that well, or in that Lake, or in that river and ask for anything and our wish will be fulfilled." This superstition is prevalent not only in India but all over the world. There is nothing wrong with giving this superstition the title of global superstition.

If there is a river or lake in the vicinity of a place where we have gone for a walk, we must have seen people throwing coins into that river and trying to fulfill their desires. I have seen people throwing money in many wells or reservoirs in this way and hoping for their desire to come true.

Maybe it is also true and it may be that someone has thrown a coin in this way and asked for crores of rupees and became a millionaire. But I have never seen such an example in front of my eyes. Alas, sir, if this is how any wishes are fulfilled, then I will sit by that reservoir for the rest of my life and throw coins and ask for things and a good life. Yes, if you have to throw a coin and you are going to ask for money, then you should throw more money to that holy reservoir and ask to make you a billionaire.

If this were true, no one in this world today would go to bed hungry. Sir, not only about sleeping hungry, no one is poor and one to ten names of the richest people in the world will be updated daily.

Coins were thrown into rivers or reservoirs by our ancestors, but as the years passed and the true scientific reason was to disappear, we today blindly imitate it only as a superstition. Maybe if we had thought a little we would have known why our ancestors were throwing coins in reservoirs?

So, let's find out the scientific reason for this too. In earlier times there was no reverse osmosis (RO), water filter, or anything like now. At that time only the water of the river or the surrounding reservoir was used as the source of drinking water. Another important point is that in earlier times coins were made from copper.

So our ancestors used to throw these coins in the water i.e. in the reservoir. They were probably aware of a scientific reason that putting copper coins in water, processing copper with water, provides the body with the right amount of copper, and copper is essential for the body. Besides, putting copper in the reservoir also destroys the harmful bacteria in it.

one person couldn't throw copper in such a large reservoir at once, so all the people were throwing coins in the form of coins for water purification. But years went by and even this scientific reason hid behind the veil of superstition. Today, even though our currency coins are made of stainless steel, we have thrown the coins in reservoirs and maintained superstition. If you throw coins in the reservoir like this, it's fine, no problem at all.

But what about the waste we inadvertently dump into reservoirs?

By doing this, will we be able to give a better future to our future generation?

A better future aside, can we give them pure water that can be drunk without a filter?

Think about it and yes not only think about it but also execute.

SEVEN

The Day of Cutting Hair and Washing the Head

> *"A belief which leaves no place for doubt is not a belief; it is a superstition"*
> *- Jose Bergamin*

For years, haircuts and head washing days have been fixed in India. Only during this day, we can cut the hair and ladies can wash their hair. This appointed day is usually a Sunday. Apart from these days, if someone cuts hair or washes one's head, they have to face those peoples who believe in superstition.

Almost all of us must have faced this superstition. This is also one of the most prevalent superstitions that have been going on for years. Yes, from now on some big cities have less of this superstition but it cannot be said that it has stopped. Once I thought that it was not appropriate to include this superstition here, but then the thought came to me that the observance of this superstition had diminished but not completely stopped.

People usually avoid getting their haircut on Tuesdays and Saturdays. As well as ladies avoid to wash their heads on Tuesday, Thursday. sometimes it is forbidden to wash the head on Saturday. All this is the result of a tradition that has been going on for years. It is believed that Saturday is Lord Hanuman's day and on this day Lord Hanuman gets annoyed if someone's hair is cut. As well as Tuesday is similarly infamous.

Many barbershops are closed on Tuesdays. Yes, even if they are not to blame, what will those people do? I don't understand why lord Hanuman gets annoyed if we cut our hair.

All this was nothing but a different method of conserving water. Years ago, people did not have borewells in their homes so people had to fetch water from wells or nearby reservoirs. In addition to this, the water coming from the corporation was also distributed in a limited number of days due to its limited quantity.

So, our ancestors used to cut their hair and wash their heads on the day when water was provided by the corporation. From this, the readers of Saurashtra will know this very well. Because even today in Saurashtra, corporations or municipalities distribute water in a few days, the rest of the days' people use the stored water for daily purposes.

There is no scientific reason behind this superstition but it can be said to be a practice or system that has been going on for years. Thus, during the days when water was distributed, people cut their hair, and women also washed their hair. Apart from this, people used to avoid cutting their hair or washing their heads due to a lack of water.

This is the way people have been following this system for years. Now due to the habit of the years the system had settled down very nicely with our life, and had become a lifestyle. Years passed and it did not take long for this practice to turn into superstition.

Today, even in the 21st century, the facility of borewell has come to the homes of the people, and water has also distributed to homes of the people round the clock. Yet today this superstition is considered in many places and is followed at the same time. Sir, no god or goddess has anything to do with washing or not washing your head, as well as cutting or not cutting your hair.

I am just talking about the logic behind a superstition. It is not wrong to follow this superstition for any reason. At this time, we should not waste water as much as possible. So that our future generation can have a better future and water.

Glossaries

- **Saurashtra**- A particular region of Gujarat known as saurashtra

EIGHT

Bathing after a Funeral

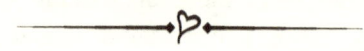

""Superstition is more dangerous than any epidemic."

- Akshay Bavda"

In our country India, if anyone dies, his body is cremated or buried, and before that rituals are performed with his body. So that his soul may attain peace and tranquility. After all the rituals are completed, the body is taken by the people and the final journey is taken out. In the end, the dead body is cremated.

People who have participated in this final journey or final action are obliged to take a bath. In the past, many funeral homes were built around the river for the above mentioned reason. This tradition, which has been going on for years, was created by our ancestors. But even this tradition did not take long to be spoilt. According to superstition, if a person does not take a bath after the funeral, the soul of the deceased haunts him. As well as his family members also face serious problems. The scientific reason for the superstition that has prevailed for so many years is quite different and nice.

The practice of hygiene and sanitation was different when this tradition was implemented many years ago. Also, the facilities available at that time were not as advanced as they are today. In earlier times, funeral homes were usually built around the river. So, they made the environment a little more conducive to the growth of bacteria. At the same time, when a person dies, the body loses its ability to fight bacteria and begins to rot. People who attend the funeral come in contact with the dead body and therefore also come in contact with bacteria, which take part in the decomposition of the dead body. This is the reason why our ancestors were told to take a bath immediately after the funeral before touching anything or anyone.

The fact that almost every part of India falls in the temperate zone also contributes to speeding up the process of decay. There are very few people who slow down the process of decomposing a corpse by laying it on an ice sheet and providing a cool atmosphere. Also, there may be scavengers around the cemetery who are usually a great carrier for bacteria.

It is not known when the tradition of bathing turned into superstition. Thus, behind every superstition there was a nice scientific reason or a noble logic. But the foundations of tradition which were built based on scientific reason, years ago are slowly being hollowed out and the building without base is being built on the foundation of superstition.

NINE

BROKEN MIRROR OR GLASS

> *"Imperfect knowledge is the source of superstition."*
>
> *- Akshay Bavda*

This superstition is believed in many places in India. Breaking a mirror or glass is considered very inauspicious. Besides, in some places it is considered inauspicious to have a broken mirror or glass in the house. Women have been using mirrors for centuries and even today they use them many times more than before. There is a mirror in the dressing table of the bedroom, so the husband and wife have equal rights over it, but even if the husband just wants to comb his hair, he does not get a place. Anyway, these are the problems of his personal life. We are here to talk about problems like superstition. The mirror was invented by a chemist in the year of 1835. Before that people used to fill water in the container and make temporary mirrors to see themselves. Then came the era of making mirrors from copper and now finally mirrors are made from glass.

Mirrors used to be very expensive in earlier times. So that earlier people used it very sparingly. Thus, in order not to damage the mirror while using it with the utmost care, one thing was spread that if the mirror is broken, it is a symbol of bad luck. After all these people make this thing so popular that even today, we can see the reflection of this thing as superstition.

Along with this, the superstition that keeping a broken mirror or glass in the house can cause harm or inauspiciousness has also taken root. There is also a common logic behind this superstition. In the past, people used to live in the dark after the evening due to lack of light in the house, only with the help of lanterns. If any broken glass or mirror falls in the meantime, it can cause serious injury. So, in earlier times people avoided keeping broken glass as well as mirrors in the house.

Yes, even today this superstition is found in many places. Thus, keeping broken glass or mirrors in the house is very risky, so instead of keeping broken glass and mirrors in the house, it is very important to dispose of it in such a way that it does not harm anyone else. But yes, there is no need to teach our children that living with this mirror or glass in the house will be inauspicious, instead, it is necessary to bring them face to face with the truth. Thus, we have to stop this superstition. If we continue to imitate blindly as before, then this superstition will continue traditionally.

TEN
No sweeping after sunset

""Scientific explanation is a factor in eradicating superstition."

- Akshay Bavda "

In our India, we may have heard many times that it is considered inauspicious to sweep garbage in the evening. But even this is nothing but just a superstition that has been going on for years. People do not know the real reason behind this superstition.

It has been believed for years that if someone cleans the house in the evening, their fate will be wiped out with that garbage. Lakshmiji gets annoyed with him and Panotiji arrives at the house. With this arrival, the clouds of danger fall on the house and it rains of unfortunate.

Thus, even if you hear such a terrible reason, you will never sweep again. But even with this matter there is a nice logic hidden. Which usually no one knows. Also, no one dares to think about this subject.

This superstition has been going on for years, so to understand this superstition we have to shed light on the times before. In earlier times in India, electricity was not available everywhere. So, people were subsisting on low-light emitting devices like lanterns. So that people had to face darkness after sunset.

Thus, after the evening spreads a blanket of darkness, if anyone sweeps trash from the house, they might lose something valuable by mistake. At the same time, in earlier times, houses were made with the help of cow dung. So that in the dark, life can be endangered by accidentally approaching a dangerous species like a snake or other venomous creature that has inadvertently preyed on a small animal like a rat.

Due to such reasons, people did not sweep the garbage after an evening in earlier times. It is not known where this truth got lost in the darkness over the years, as well as the logic turned into superstition. Even today we have electricity and light facilities and still some people do not clean the garbage in the evening time.

Glossaries

- **Lakshmiji** – Goddess of Money
- **Panotiji** – Goddess of Bad luck

ELEVEN

Crushing a snake's head after killing

> *"Superstition vanishes before the truth."*
> *- Arnobius*

Many people die of frightened just by the name "snake". The main reason is centuries of superstition and false rumors. I am not saying that there is no need to be afraid of snakes. But we need to get some information about it. Due to lack of information, even today people tremble at the mere name of the snake.

There are about 270 species of snakes in India, out of which a total of 60 snakes are venomous and the rest are non-venomous. In percentage terms, about 22% of snakes are venomous, except non-venomous snakes.

Due to the lack of this information, years ago our ancestors used to kill the snake whenever they saw it. Then whether it is poisonous or non-venomous, it dies because it belongs to the snake species. Although this creature has been innocent for years, it has been subjected to atrocities. Usually when a snake attacks someone it does so in self-defense. But since people are unaware of this, they end the snake's life just by seeing it.

There is also a superstition associated with snakes that a snake should be beheaded after being killed. If this is not done then the snake makes a picture of the one who killed it even after death and then takes revenge. This is nothing but superstition. The scientific fact behind this is that snakes are included in cold-blooded animals. Even after the death of such animals, some of their organs remain functional. As we have often seen, when a lizard's tail is cut off for some reason, it bounces around for about 20 minutes. Cold-blooded animals usually have these types of organs.

The organs of cold-blooded animals remain functional for some time after death or even after they have been released from the body. If the snake is killed but its head is not crushed, it can harm others around it. So, in earlier times people used to kill the snake and crushed its head.

There is a reason behind this superstition. Even now, there are many cases of snakes in the house. We are increasing the forests of buildings day by day. So that the habitat of such animals is being stripped away, the circumstances of frequent visits arise.

This is a request to everyone who is reading this book that whenever, you see a snake, please do not kill it. Keep a proper distance from it, and report it to the snake catcher or the forest department. Just keep an eye on where it goes until those people arrive so it's easy to catch. Do not do any work that provokes the snake, from provoking acts the snake may come under compulsion and attack for self-defense.

TWELVE

No nail cutting and shaving after sunset

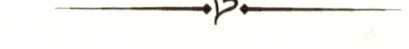

> *"Science is the great antidote to the poison of superstition."*
>
> *- Adam Smith*

This superstition is also one of the prevalent superstitions. Usually we have often heard from our grandparents that we should not cut our nails or shave after evening time. If this is done then something bad happens to the person, as well as the wealth in his house does not last. According to many superstitions, the reason behind this superstition is the lack of electricity. Years ago, people used to avoid cutting their nails and shaving in the evening due to darkness and lack of electricity.

There is darkness in the house due to lack of electricity and if at such a time a person sits to cut his nails and there is a danger of getting injured by scissors due to darkness. Similarly, if a person shaves, the risk of getting injured by a blade increase. At that time there were no Gillette's razors with diamond coating blades or nail cutters like now. So, people avoided cutting their nails or shaving in the evening.

Thus, every generation was instructed not to cut or shave their nails in the evening. But no one gave a true reason so people blindly believed what their elders said and gave us a superstition in the result. This is still prevalent today. People are still hesitant to cut their nails or shave in the evening.

THIRTEEN

Forbidden to plucking flowers after sunset

"Just as darkness promotes the growth of reptiles, so, on the contrary, the creatures of superstition promote the growth of darkness."
- Lady Blessington

Flowers are considered very sacred in India, as well as used in every occasion or during worship. In India, the flowers are used extensively to worship god. Apart from this, the flower attracts the attention of everyone because of its various captivating colors and its beauty.

We have all heard from our grandparents many times that "flowers should not be plucked after evening time". If this is done then God is offended and it is considered inauspicious. Many people still believe this and avoid plucking flowers in the evening.

But there is a logic behind this superstition that no one has ever thought or even tried to think. The main reason for not picking flowers after the evening is that usually in the evening all the creatures have returned to their home. Besides, small plants and trees attract and provide shelter to many small and large creatures, reptiles, and venomous creatures due to the environment around them.

Due to the lack of electricity in the past, if a person goes to pluck flowers in the dark in the evening, they may come across poisonous insects. So, our ancestors forbade plucking flowers after evening. Time passed and this logic also turned into superstition. Even today we believe in this superstition.

FOURTEEN

SLEEPING WITH AN IRON UNDER A PILLOW

"The sun rises behind people's backs and they are worshiping darkness."
- Robert G. Ingersoll

My younger sister had a lot of troubles like having nightmares at night as well as falling asleep in the middle of the night. Almost every night she had dreams of ghosts. One day all of a sudden, my mom said this to my grandmother. Grandma replied that if she put an iron-edged object under the pillow she would stop having bad dreams. According to grandma, Mom did this experiment and the result was very shocking. My sister stopped having nightmares. I saw this phenomenon when I was very young but my brain kept asking me questions.

How is this possible? Why did bad dreams stop coming from just having an iron object?

Many such questions arose in my mind but no one could answer me. I asked a lot of people, all of them gave different reasons but no one could satisfy me.

Shortly afterward I was watching a program on the Discovery Channel when I came across information about "Flinders Bars" during that episode. Then did I get the idea that our brain works on electromagnetic waves. So, if an abnormal magnetic field is generated in the room around you or in the place where you are sleeping, it disturbs the working capacity of the brain in its sleep state which is out of our control, and dreams of ghosts are realized.

"Flinders bars" are used under the ship's magnetic compass. Which are soft iron bars used to correct errors in the magnetic field, Flinders bars fix the compass of ships after placing, and show the correct direction.

Thus, this superstition has been going on for years but no one knows why iron is put under the pillow. Yes, people have a solution to the problem but do not know the exact reason behind it. By placing an iron object under the pillow and sleeping, it acts as "Flinders bars" and eliminates the abnormal magnetic field created around it so that the disturbance to the brain is reduced and the person can sleep well.

FIFTEEN

KEEPING HEAD NORTH FACING DURING SLEEP

"What we don't understand we can make mean anything."
- Chuck Palahniuk

This is also one of the most prevalent superstitions in India. It has been believed in Indian culture for centuries that you should not keep your head facing north while sleeping. Because in that direction, when a person dies ancestors keep dead bodies head towards north.

So, if a person sleeps with his head facing north, his elders stop him from doing this thing. And if a person asks why this should not be done, the only answer is that the head of the corpse is always kept in that direction.

Yes, this superstition is very useful but it is necessary to know the proper scientific reason behind it. We all know that our earth is a big magnet, there is a magnetic field around the earth. Along with this the brain also works with the help of electromagnetic waves. Thus, it is very important to consider the direction of the poles of both the magnets when they come together.

So, if we sleep with our head facing north, we disturb the earth's magnetic field and our brain's magnetic field. Therefore, if we sleep in the north direction for a long time, it will also have a bad effect on our blood circulation. So that the risk of hemorrhage or tumor increases.

Our ancestors were aware of this truth so they kept preventing us from sleeping to heading North. But years went by and the scientific reason disappeared somewhere and superstition was born without a reason.

SIXTEEN

MENSTRUATION

> *"The Devil divides the world between atheism and superstition."*
> *- George Herbert*

The best creation of God is "woman". In Indian culture, women are considered as goddesses. Some people respect women while some people don't understand women at all. From ancient times the woman has traveled a long way, in which the woman is the "goddess" and the woman's intellect is lesser than men. Thus, it is said that woman is the best creation of God. So, it means that we, the trivial human beings with blackheads, also find flaws in the creation of the Creator.

For a long time, women have been suffering a lot of atrocities. As jokes are made about women, people write books to understand women. But I just don't understand why we have to understand women. How many people understand the theory of relativity of physics? I don't know how many people asked questions against it? Accepted peacefully or not? Thus, woman is also a complex creation of God, if you keep trying to understand it, it's worthless. Many people came to try to understand the woman but they got only failure as a result. Without trying in vain to understand the woman, there is no need to understand, just accept her with love just like she is.

After reading this, it must have happened to many that what is this? We always tried to understand the woman but no one tried to understand the female body? About 80% of men in this world will not know the reason why a woman menstruates.

Yes, those people will also believe in the superstition that "a woman is unclean during menstruation, she should not go to religious places and should not even enter the kitchen" and will blindly cooperate with many such superstitions. No one thought that from where this superstition came from.

Menstruation is a process associated with a woman. This process starts from the ovary, the egg is created in the ovary. The egg slowly moves through the fallopian tube to the opening of the uterus. As this progress, the size of the egg also increases. At the same time, change takes place in the uterus to establish the egg. This change causes the wall of the uterus to grow and become thicker. Now this egg matures over time and if a woman has sex during this time and this egg comes in contact with sperm it can be fertilized. If fertilized, the fertilized egg is implanted in the uterus. But if not fertilized, the egg slowly begins to melt and the walls of the uterus return to their original form. With the bleeding, the unborn egg comes out through the vaginal tract. This bleeding occurs in almost every woman at different times from 3 days to a week.

Now let's talk about superstition, in India years ago a woman who was menstruating was considered unholy.

They were also forbidden to touch anything or go to places of worship. The scientific reason behind this is that before the invention of sanitary pads in 1888, women used to wear normal clothes during menstruation. So that there was a fear of spreading harmful bacteria. Menstruation is mainly lumping of protein so due to high protein there is a very high risk of bacterial growth. So, at that time if the woman touched anything during this time the risk of bacterial infection was increased. Apart from that, menstruation is very painful and the woman needs a lot of rest during this time due to hormonal changes. So, in earlier times a woman was not allowed to work.

But time passed and this true scientific reason also vanished, and people began to consider women unholy in this time. Even today many people believe in this superstition and consider women unholy during this time. There can be no defect in any of the creations of God and this process is also indispensable for giving motherhood to a woman.

SEVENTEEN

ECLIPSE

> *"Superstition is the religion of the feeble minds"*
> *- Edmund Burke*

The astronomical event "Eclipse" is considered very ominous in India. Even today, in the twenty-first century, many people consider the eclipse to be ominous. While the elderly people are still running away after hearing the name of the eclipse.

This is also a superstition that has been going on in India for centuries. At the time of the eclipse anyone should not go out of the house or watch the eclipse. People believe that evil spirits, as well as ghost powers, are at their peak at this time. Also, many people consider the eclipse to be the menstrual cycle of the earth. Many such superstitions are associated with the eclipse.

But people do not know how this superstition came into practice. There is also the science behind this superstition which our sages knew and they must have told the people and like every superstition this time too the scientific reason has disappeared and if superstition remains then let us know the scientific reason behind this superstition too.

An eclipse is an astronomical event. There are two types of eclipses, lunar eclipses and solar eclipses. We all know that the earth revolves around the sun and the moon revolves around the earth. Thus, during this orbit, when the earth, sun, and moon come in a line, an eclipse takes place corresponding to the position of the moon. A solar eclipse occurs if the moon comes between the sun and the earth. A lunar eclipse occurs when the earth comes between the moon and the sun.

Scientists have proven that during the eclipse, the emission of ultraviolet rays from the sun is very high. So that if a person comes out during the eclipse that person may be harmed by harmful rays. Also, if seen with the naked eye, a person's eye can be damaged and the person can also lose eyesight.If a pregnant woman goes out during this time, her fetus can also be affected by bad rays. Thus, for such reasons, it was forbidden to go out during the eclipse but peoples made it a complete superstition and even today this superstition continues in this twenty-first century.

EIGHTEEN
Wearing silk during worship

"Tradition is a guide, not a jailer"
- W. Somerset Maugham

Pooja means to take the blessings of the goddess or deity we believe in. Worship is of paramount importance in Hindu culture. Worship is also done in the temple and everyone does it at home. A small temple is installed in each person's house and is also worshiped. The custom of wearing silk during worship has been going on for years in Indian culture. Mostly during pooja, the woman wears a silk sari and the man wears a silk dhoti and Khesh. Our great ancestors have followed this tradition for centuries, and have also passed it on to future generations.

This tradition that has been going on for years is very beneficial and science is also hidden behind it. "Silk" is considered an excellent fabric. Silk carries a very good amount of electromagnetic waves. At the same time due to friction between silk cloth and human body static electric current is generated. If a person wears silk clothes and worships, the positive energy generated during the worshiping as well as the electromagnetic energy enters the body of that person. Due to the good energy carrying capacity of silk, more and more positive energy is given to the worshiper so that the person experiences instant peace of body and mind. In addition to being very comfortable in wearing silk clothes, the focus of the worshiper can be very good.

Keeping in mind these major scientific reasons, silk is worn especially during worship in India. So, we have to maintain this tradition, it is our duty to make this science clear to the coming generation as well.

Glossaries

- **Pooja-** Worship
- **Sari-** Indian traditional wear for women
- **Dhoti-** Indian traditional bottom wear for men
- **Khesh-** Indian traditional top wear for men

NINETEEN
VERMILION

""An Examination of Indian Vedic doctrines shows that it is tune with the most advanced scientific and philosophical thought of the West."
- Sir John Woodruff"

The word "Sindoor" is considered very sacred in Hindu culture. After getting married, every woman wears a vermilion on her head which is called "sentho" in Gujarati. For the first time any woman uses vermilion during her marriage when her husband completes all rituals of marriage and in the final stage fills her "sentho". According to Hindu culture, in a marriage, the man completes the sentho of his wife and the mangalsutra is worn and the marriage is completed with the promises of the saptapadi.

After getting married, every woman fills her sentho with vermilion. Most people don't know that

Why does a married woman complete Sentho from Sindoor?

Why is a virgin girl strictly forbidden to touch the vermilion?

Why can't a widow complete her sentho or touch the vermilion?

To understand each of these questions as well as Hindu culture, we need to cultivate a scientific approach. Vermilion is made mainly from a mixture of three substances turmeric, lime, and mercury in metallic form (mercury). This mixture applied by a married woman i.e. vermilion is very beneficial. The mercury in vermilion controls a woman's blood pressure. And simultaneously increases sexual impulses. Along with this it provides peace of mind as well as helps in relieving fatigue.

The use of vermilion has decreased in married women over time. Also, some women use vermilion in very small quantities, which means they follow both tradition and fashion. In this age of fashion women now only do sentho for a false conscience so that women apply vermilion in a very small area. But if the vermilion is not just applied formally, it is very beneficial to apply it from the forehead to about three or four inches to get the maximum benefit. This is because there is a pituitary gland in this area that regulates emotion and sensation.

The use or touch of vermilion greatly increases sexual arousal. Vermilion is not allowed to touch a widow or a virgin because it increases sexual arousal. We are blessed by the ancient Indian culture and our ancestors who also achieved important achievements in the field of science and have been given to us as a legacy.

Glossaries

- **Sindoor**- Vermilion
- **Sentho**- Applying vermilion (Indian tradition)

- **Mangal sutra**- sacred jewelry for Indian married women
- **Saptapadi**- Giving seven promises to each other to solemnize Indian marriage

TWENTY

Festival of Deceased relatives

> *"There has been no more revolutionary contribution than the one which the hindus(Indians) made when they invented zero."*
> *- Lancelot Hogben*

This "Shraddha Paksha" means the month of the forefathers. We all know very well about Pitru and Pitrumas. Pitrumas means from the worship of Peepal to feeding crow. You may have the next statement that I forgot the Dudhpak. No, my dear friends, I have not forgotten Dudhpak. Shraddha is not possible without Dudhpak, and we are talking about the whole month, so the whole Shraddha is incomplete without Dudhpak.

So, let's get to the point now, this tradition that has been going on for years i.e. "Shraddha" and "Shraddha Paksha" is very well celebrated in Hindu culture. During Pitrumas, people celebrate the day of death of their deceased relatives in their home by making puri, vegetables of potato, curry, rice, lentils of mug, and dudhpak. On this day, all the people of the house get up early and perform pooja and donate. As well as doing good deeds for the souls of all their dead loved ones to get peace. After completing all the rituals, one of the men of the house goes to the roof of the house or on the terrace and feeds the crows a mixture of puri and Dudhpak. And on such a special day the stray dogs are fed biscuits or milk-bread, or shira.

If we look at this whole month from a scientific point of view, it will be understood, and we will also get proper solutions to the various questions that arise in our minds. Such as

Why do shraddhas come especially in this month?

Why are only crows to be fed?

Why are dogs also fed biscuits or milk or shira?

From a scientific point of view, Shraddha Paksha comes in the month of bhadarwah. The full respectful month is known as the pitrumas. Bhadarwah is the month of mating for both dogs and crows. Dogs are fed during the pitrumas so that the dog gets full nutrition during pregnancy. Since crows also mate in the month of bhadarwah, the female lays eggs in the month of bhadarwah. As well as hatches it and in the month of bhadarwah and baby crows also born from the eggs in this month. Our ancestors considered the month of bhadarwah as the pitrumas to provide adequate food and nutrition to these chicks.

Now in 21st-century people like us may wonder why our ancestors were trying to save the crow species. Even so, crows make irritating noises the whole day outside the house. There is a scientific reason behind it. Crows are the only species that eat the seeds of trees like Banyan and

Peepal called "Teta". During the digestion of crows, this teta is processed and able to grow, and where crows defecate, the chances of growth of banyan tree as well as peepal tree are greatly increased.

The Banyan tree has been given a special place in medicine. Banyan tree is used in making many medicines as well as herbs. The peepal, on the other hand, is a tree that provides oxygen for about twenty-four hours and purifies the atmosphere. So, the existence of banyan trees and Peepal trees is very important for the protection of nature and at the same time the existence of the species of crows responsible for the existence of these trees is also very important.

Pray in the month of Bhadarwah for the protection of nature and to save the species of crows from extinction.

Glossaries

- **Shraddha Paksha**- Worshiping time for deceased relatives
- **Dudhpak**– Indian dessert made from milk and rice.
- **Shraddha**– worshiping for Deceased relatives
- **Puri**– Indian bread
- **Shira**– Indian dessert
- **Bhadarwah**– Month according Indian calendar
- **Teta**- Seeds of Banyan and Peepal tree called in Gujarati language

TWENTY-ONE
Cow dung

"Vedas are the most rewarding and the most elevating book which can be possible in the world."
– Arthur Schopenhauer

The cow is an animal that is revered, we have given the cow the title of mother. Cow's milk is thus very beneficial for everyone. Along with this, cow dung is also full of virtues. Many of us may have noticed that people in the villages did not let the cow dung go to waste. Collect the cow dung and press it with both hands and shape it round and flat and let it dries on the wall. This dried manure was then used as fuel in the Chula.

In addition to being used as fuel, the ancients used to smear manure in their homes. That is, they used to apply a layer of dung on the ground in the house as well as on the adjoining wall. Besides, the villagers used to dilute the cow dung with water and sprinkle it near the porch. After reading all this, the word "shit" must have come out of the mouths of some people. But such people will be very surprised in a short time, knowing the science behind this tradition.

The scientific reason behind this is that cow dung is an excellent source of methane. Bacteria in cow dung helps to produce methane, which is a by-product of the oxidation process of these microorganisms. And when methane chemically reacts with sunlight in the presence of air it is converted to formaldehyde. Formaldehyde is an excellent disinfectant. So that it kills all the harmful bacteria and viruses around. So that a pure atmosphere is found.

Methane + oxygen + sunlight = formaldehyde

Thus, our ancestors were very intelligent and very knowledgeable in the field of science. It seems to us that smearing this dung is very dirty as well as unusual but over the years our ancestors have been using science in this way and various other ways.

Glossaries

Chula– Ancient Indian stove basically use cow dung as a fuel.

TWENTY-TWO
NAMASKAR

""There is no book in the world that is so thrilling, stirring, and inspiring as the Upanishads."
- Max Muller"

Today we greet and shake hands with anyone we meet. But years ago, our ancestors used a different way to greet. Some people still follow this old tradition and welcome anyone with joining both hands together. At present, the method of this salutation seems to have become extinct. All meet and shake hands. Our old tradition was very good. We can protect us as much as possible from avoiding contact with people in an environment like coronavirus.

Our extinct tradition may be re-established due to the coronavirus. The speed at which we are moving towards Western culture is much higher. We have to put a full stop to this movement because the people of the western culture are overwhelmed by the greatness of our Indian culture. But it may be too late for our people to realize that.

Our old tradition of welcoming with a joining pair of hands together was very good. There is a scientific reason behind it. Whenever we put our hands together, the fingers of our hands touch each other. With this touch, it is also a little pressed. The pressure points of our eyes, ears, and brain are located on the tip of our fingers. Pressing from this point activates the eyes, brain, and ears nerves so that we remember the visit with that person very easily and for a long time.

Another great advantage of this tradition is that greeting from a distance and not touching it reduces the chances of harmful bacteria as well as infectious diseases. Thus, following the traditions of our great ancestors will be more beneficial for us than following the traditions of Western culture.

Glossaries

Namaskar- Greeting by joining two hands

TWENTY-THREE
NAVRATRI

"Gravitation was known to the Hindus (Indians) before the birth of Newton. The system of blood circulation was discovered by them centuries before Harvey was heard of."
- P. Johnstone

As soon as the name of Navratri is mentioned, Gujarati are filled with a lot of enthusiasm and happiness. The feet begin to move in rhythm to take garba and the mind becomes very happy and falls in remembrance of Garba.

Navratri is the most important festival of the Hindus, which is celebrated with great fervor and exuberance all over the country. Traditionally associated with Goddess Durga and worshiped in her nine incarnations, the festival is celebrated with great fervor in North India, West Bengal, as well as in the central and western regions of the country.

Some people fast all day in Mataji's faith, most people fast only in the first and last days, and some people abstain from consuming only alcohol, onions, garlic, or non-vegetarian items. Another important thing is that the list of dishes to be eaten during this fast is different. Because there is a strict ban on certain sets of food items during this festival.

There are just two types of beliefs behind this custom (tradition). One is the traditional belief i.e. what people believe and the other is the scientific belief. For religious reasons, fasting is a way to get closer to the Almighty. Many cultures believe that spiritual purification from renunciation leads to simple willpower. It is also seen as a way to pursue virtues like self-discipline and futility and to save oneself. So those who fast give up their regular diet and turn to light food items as a way to renounce and draw closer to God. Many people also prohibit consuming water on fasting days. This is known as the "no-food-no-water" fast Nirjala fast.

Traditionally in Hinduism, the consumption of alcohol and non-vegetarian food is considered inauspicious and unholy, there is solid science behind it. Navratri is celebrated twice a year and if you notice, every time it comes during the change of seasons. From an Ayurvedic point of view, eating foods like meat, grains, alcohol, onions, garlic, etc. attracts and absorbs negative energies. Our body's immune system is very low due to the change of seasons during this period. As a result, the chances of getting sick are greatly increased so this type of food should be avoided.

The body needs enough time to adjust and prepare itself for the changing seasons. These nine days were marked as a period when people would cleanse their body system by fasting, avoiding excessive salt and sugar, meditate, gain a lot of positive energy, gain confidence and increase self-determination (fasting is a means to improve our willpower and self-determination) And finally the body will be ready for the challenges of the changing seasons. Therefore, it can be said that our ancestors were smarter than us and had a scientific approach.

Glossaries

- **Navratri**- Nine days sacred festival of India
- **Garba**– Typical dance for celebrate Navratri
- **Mataji**- Goddess
- **Nirjala fast**– Fast without consuming Food and Water

TWENTY-FOUR
WEARING A BANGLES

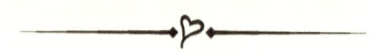

> *"We owe a lot to the Indians, who taught us how to count, without which no worthwhile scientific discovery could have been made"*
> *- Albert Einstein*

In Hindu culture, women have been wearing bangles for centuries. The bangles have become ornament for Indian women. We will also see every woman around us wearing bangles.

Women have made this jewelry a part of their lives. Every married woman is required to wear a Bangles. Because it is considered the sign of a bride.

This tradition is one of the traditions that has been going on for years. There is also a scientific reason behind following this tradition.

According to our Ayurveda, women's bones are much weaker than men's. Bangles were usually made of gold and silver in earlier times. The qualities of these noble metals are unconventional. These metals have properties that can absorb the energy produced by the body. Bangles made from this metal absorbs the energy produced by the body and carry it back into the body instead of letting it go to waste. So, with the help of this energy, the functions of the body as well as efficiency are increased.

At the same time the mutual friction in the Bangles, as well as the constant friction with the wrist, makes the vein healthy for better circulation of blood in the part of the wrist. In earlier times vascular science was used to diagnose major diseases so that the better the circulatory system, the sooner disease was diagnosed and sooner was resolved.

For these reasons, in earlier times women used to wear Bangles as ornaments. So, Bangles made of metal like gold and silver prove to be beneficial for the body. But even today this tradition continues but the only difference is that gold and silver Bangles have been replaced by glass and plastic Bangles. Which can only be worn as an ornament to enhance beauty but to no avail.

Even today, if a woman wears a Bangles made of metal like gold or silver, she can still get benefits. But these Bangles made of glass and plastic can be used only for fashion.

TWENTY-FIVE
HENNA

> *"After the conversations about Indian philosophy, some of the ideas of Quantum Physics that had seemed so crazy suddenly made much more sense."*
>
> *- W. Heisenberg*

Indian weddings are known for their many rituals. The beauty of Indian weddings comes from the myriad traditions associated with special celebrations. Indian marriage is incomplete without dance, music, lots of rituals and jokes, fun, and laughter. Also, Indian weddings are not a one-day event. Pre-wedding ceremonies begin before the wedding and sometimes it can be a week-long celebration.

It is a common belief that the darker the color of the Henna on the bride's hands, the more it will be preferred and loved by her husband and mother-in-law. However, the importance of applying henna at the time of marriage is not limited to emotions and beliefs. Although these beliefs make the use of henna a much anticipated and enchanting tradition, their real reason is very important, which is sometimes forgotten in the present times.

In addition to lending color to the hands, henna is a very powerful medicinal plant. Marriages are stressful and often lead to headaches and fevers. As the wedding day approaches, there is a feeling of excitement as well as panic in the woman. The use of henna can relieve excess stress as it cools the body and protects the nervous system from stress.

This is the reason why henna is applied on the hands and feet, besides, being a very excellent antiseptic agent, henna can protect couples from viral diseases. Such diseases are quite undesirable even before marriage and this herb can provide strong shield-like protection. It improves blood circulation in the body and enhances general health. The early practice of applying henna in ancient times was started to protect the couple from diseases and keep them healthy at the time of marriage.

The Henna that is applied during Indian weddings is not just a simple paste of Henna powder and water. Eucalyptus oil, a little clove oil and a few drops of lemon are added into Henna. This oil not only helps to darken the color of henna on the body, but also enhances the benefits of henna and makes the paste very medicinal. The best part is that the smell, the beautiful rich color, and the health benefits that henna gives act as a powerful sexual impulse enhancer. Also, the color and smell of henna lasts for days, so it boosts romance in the early days of the wedding.

TWENTY-SIX

EAR AND NOSE PIERCING

> *"They were very advanced Hindu astronomers in 6000 BC. Vedas contain an account of the dimension of Earth, Sun, Moon, Planets, and Galaxies."*
> — *Emmeline Plunrate*

"Ear Piercing" It is an Indian tradition that parents pierce their daughter's ear at a very young age when the skin is very soft. Almost every woman has her ears pierced. As well as nose piercing has also become our tradition. At the same time, two additional items were added to the woman's jewelry.

First of all, if we talk about ear piercing, in Ayurveda it is believed that piercing a woman's ear protects against diseases like a hernia. Ear piercing increases intellectual capacity as well as decision-making ability. At the same time ear piercing also controls a woman's menstrual function.

In earlier times, when a woman was eligible for marriage, her nose was also pierced. But nowadays even if a woman is not worthy of marriage, her nose is pierced and she wears a nose ring for fashion.

But in earlier times, a woman's nose was pierced only when she was eligible for marriage. Yes, some people didn't follow this thing so they pierced a woman's nose even at a young age. Even in earlier times, child marriages were performed in India, and nose piercing was done at an early age.

There is also a scientific reason behind our tradition of piercing the nose. In India, the left side of the nose is generally considered to be the ideal position for piercing. According to Ayurvedic mythology, the place created by the piercing of the nose is associated with the female reproductive organs. It acts as an acupressure point. It is generally believed that a woman who has her nose pierced on the left side, feels less pain during childbirth and also has less menstruation. Wearing a nose ring makes the birth process easier.

Thus, it can be said that our ancient culture was so great that there is a beautiful scientific reason behind every religious tradition.

TWENTY-SEVEN
Toe Ring

> *"India will teach us the tolerance and gentleness of mature mind, understanding spirit and a unifying, pacifying love for all human beings."*
> *- Will Durant*

This jewelry is usually a part of a married woman's jewelry. Yes, foot rings, toe rings, are identified in different ways. This ring is worn on the side of the toe in the foot of any married woman. This ring is usually worn only by a married woman. There is also a scientific reason behind it.

This ring is made from silver. According to Ayurveda, the nerves in the finger on the side of the toe of a woman are directly connected to the uterus and heart.

So, wearing this ring increased the blood circulation of uterus and heart. And this strengthens the walls of the uterus and makes it suitable for establishing a baby. As well as the nerves of the fallopian tube is also located in this finger and the ring worn here also causes regular menstruation. So that the fetus can be conceived very quickly.

In addition silver is an excellent conductor, it absorbs good energy from the earth and spreads it throughout the body through the feet so that positive energy is transmitted thoroughly the body.

There will be many of us who do not know the scientific reason behind this tradition. This is the greatness of our Indian culture that with every small tradition comes with a big and beneficial science.

TWENTY-EIGHT
Tilak

> *"From the Vedas, we learn a practical art of surgery, medicine, music, house building under which mechanized art is included. They are encyclopedia of every aspect of life, culture, religion, science, ethics, law, cosmology and meteorology."*
>
> *- William James*

In India, women are seen wearing bindi on their foreheads. But most of these people in India and around the world do not know the reason and follow only in the name of custom or tradition. There is no traditional belief attached to this custom. It is the only material that is used to add beauty to women. The red bindi is also said to give a certain beauty to women when they wear it. Which corresponds to the main points of the body (concerning the nerves). They are considered to be the centers of cosmic energy connection of the body. This principle forms the basis for many practices such as traditional healing, Reiki, and primarily yoga.

Ajna chakra displays a person's intelligence. As well as any bad or negative thoughts, as well as energy, is transmitted only through this chakra. When Bindi is placed in this space, the negative energy entering the body is blocked with the help of Bindi.

The traditional bindi is made from kumkum which is turmeric and saffron. Turmeric has many uses in culinary and health sciences and cosmetics. This excellent blend made from turmeric and saffron also has a different scientific significance. Kumkum (a mixture of turmeric and lime) or sandalwood was traditionally used to mark the bindi. Kumkum is hygroscopic and can greatly help remove excess water from one's head. The antibacterial properties of turmeric are also helpful. Sandalwood is famous for its 'cooling' properties. For a warm tropical country, the use of sandalwood on the forehead (nerve center) helps to cool the whole system of a person.

I realize that the explanation of this tradition is based on chakras, it has not been scientifically proven. All I can say is that the results of the previously mentioned techniques like yoga and Reiki have been proven and they have originated not only in India, but in different parts of the world.

Glossaries

- **Bindi–** A decorative mark worn in the middle of the forehead by Indian women
- **Ajna chakra–** one of the chakra located in middle of forehead according to Indian tradition
- **Kumkum-** Mixture of turmeric and lime
- **Reiki-**Technique of energy Healing

TWENTY-NINE

CLOSING THE EYES OF THE CORPSE

> *"I have noticed even people who claim everything is predestined and that we can do nothing to change it, look before they cross the road"*
> — *Stephen Hawkings*

Death is a bitter reality of life. The death of the one who is born is certain. When a person dies, his eyes are immediately closed. And it is believed that if this is not done then the soul of the deceased does not get peace. So that even after death soul wanders and annoys the family members.

It has been going on in our country for years that the eye of the deceased is closed as soon as possible. This phenomenon must have been witnessed by almost everyone. As soon as a person dies, his eyes are closed. Even if someone has not seen this in front of their eyes, such scenes must have been seen on the television. There is also a nice scientific reason behind this tradition that has been going on for years. According to a survey, about 63% of people die with their eyes completely closed, while the remaining 37% of people keep their eyes open after death.

According to the structure of the eye, the eye is always wet through the tear gland. But the tear gland of a dead person is not able to keep the eye wet. So that it does not get enough water to get wet.

So, if the corpse's eye is not closed as soon as possible, the muscles of the eyelid will become very tight. This will make it very difficult to close the eyes. And if the eye is not closed for a long time, the water in the eye will also dry up and as a result the face will start looking very scary.

This is the reason why whenever a person dies, his eyes are immediately closed. This was not originated only in India, but in different parts of the world.

THIRTY
HOLY BASIL

> *"Superstition is a curse for society"*
> *- Akshay Bavda*

We have seen people worshiping Tulsi as a goddess. But most of these people in India and around the world do not know the reason, why do we worship Tulsi? And just following the crowd in the name of custom or tradition.

Almost all of us will have a Tulsi plant in our house and we will all see it is worshiped daily. The basil plant is considered very sacred. With which many religious stories are connected. A different significance of Tulsi is described in these religious stories. There is also a superstition associated with this worship tradition. Many people believe that basil leaves should be swallowed instead of chewed. Because we have given the title of Goddess to Tulsi.

There is also a scientific reason behind this tradition and superstition. Scientists have proved that the Tulsi plant provides oxygen round the clock. Apart from this, many other important qualities of Tulsi also make it worthy of worship.

The chemical composition of basil is very complex. It also contains many nutrients and useful chemical compounds. Rich in these compounds and nutrients, Tulsi is mainly used as a medicine.

We have all done a very popular experiment of Tulsi as a medicine. Whenever we have a cold, drinking a decoction of Tulsi gives relief in the cold. Scientific experiments have shown that basil is very useful in treating many diseases like cancer, heart disease, arthritis, diabetes. Along with this, Tulsi also helps in controlling stomach related diseases and problems like gas, acidity, etc. Basil is also very useful in controlling high blood pressure and cholesterol.

Thus, Tulsi is an excellent plant with antiviral, antifungal, and antibacterial properties. Tulsi also contains a small amount of a chemical called arsenic. So, eating Tulsi may damage the protective enamel on the teeth so eating Tulsi by chewing makes the teeth turn yellow.

Thus, our ancestors knew the medicinal properties of Tulsi very well. So, the tradition of worshiping Tulsi has been going on in India for years. Our ancestors were not unaware of the damage caused by chewing Tulsi. So, they said that for years people stopped chewing Tulsi and due to the beliefs associated with religion, the superstition that Tulsi should not be chewed came into existence and this superstition is still going on in many places.

Yes, basil leaves should not be chewed. But we should not give any religious reason behind it to our next generation. But we should tell the truth that we cannot chew it because of the arsenic in Tulsi. So that our next generation can get rid of one more superstition.

Glossaries

Tulsi- Holy Basil

THIRTY-ONE

DIWALI

""Ancient Indian culture is the birthplace of science."
- Akshay Bavda"

Diwali is a festival that is loved by almost everyone, of any age group. Diwali or Deepavali is the festival of light. It is said that Lord Rama completed his vanvas on this day and returned to Ayodhya after killing Ravana. From that day till today Diwali is celebrated as a very sacred festival in Hindu culture.

In Diwali, people celebrate Diwali by lighting lamps and cleaning the house and decorating it in various ways. Along with this, the kids enjoy the festival with fireworks.

Thus, traditionally this festival has been celebrated with great zeal and exuberance for years. There is a very important religious reason behind this tradition but at the same time a scientific reason is also responsible.

From a scientific point of view, Diwali usually comes in October or November. During this period the monsoon is almost complete and the winter season has begun. Due to the rains during the monsoon season, many houses are slightly humid. Humidity can increase the infestation of bacteria in the house, as well as increase the chances of spreading the disease in the house. Filling water in open waste or containers also increases the nuisance of mosquitoes. So, the chances of getting diseases like dengue, malaria, or chikungunya are greatly increased.

Everyone cleans their house before Diwali. So that the garbage in the house is cleaned, the moisture is removed from the places where moisture is accumulated by keeping the goods, lying in the damp places. Along with this, the smoke produced by the firecrackers also kills insects like mosquitoes.

Thus, science is also very well connected with Diwali, the festival associated with the mythological Ramayana. Celebrated in the joy of Lord Rama's return to Ayodhya, this festival marks the end of the rainy season and the beginning of winter.

Glossaries

- **Diwali** – Indian festival of Light and positivity
- **Deepavali**– Diwali is also known as Deepavali
- **Vanvas**– Lord Ram spent 14 years in Forest this period known as vanvas (According to Ramayana)
- **Ayodhya**– Kingdom of Lord Ram (According to Ramayana)
- **Ravana**- King of Lanka and enemy of Lord ram (According to Ramayana)
- **Ramayana**- Indian mythology

THIRTY-TWO

FASTING

""The Ancient Indian Culture is a Complete Interpretation of Lifestyle "
- Akshay Bavda "

One of the traditions that has been going on for years is "fasting". Almost every one of us must have fasted at some point. Fasting is frequently observed in Indian culture. During this fast people stay hungry according to their ability. At the same time, they worship the goddess or deity for whom they have fasted. We have maintained this tradition very well even today. People still fast on Mondays, Tuesdays, Thursdays, and Saturdays at different times according to their convenience. While some people are seen fasting for two or three days in a single week.

There is also a scientific reason behind this tradition. According to the great Ayurveda of India, the structure of the human body is made up of 80% liquid and 20% solid like the earth. So that the gravitational force of the moon affects the human body which results in disruption of digestion. As well as the toxins in the food that we take unknowingly, it also accumulates in the body. Fasting is very necessary to overcome all these kinds of awkward situations.

Fasting removes toxins from the body. Fasting also reduces the excess acid in the body. Also, with the help of regular fasting, the chances of getting serious diseases like diabetes and heart disease are reduced.

Our Ayurveda is great. Years ago, our sages also achieved many achievements in human physiology. Fasting has been important in Indian culture for years. Perhaps we are unaware that according to research done in Japan, fasting can cure major diseases of the body like cancer, heart disease, etc. The Japanese cell biologist Yoshinori Ohsumi, who made this discovery, received the world's largest so-called prize, the Nobel Prize. This man demonstrated the benefits of fasting and won the Nobel Prize in 2012. Benefits from fasting has been told in Indian culture for centuries.

It is a matter of pride that those who used to consider Indian culture as a joke are now attracted to Indian culture. At the same time, it is a pity that Indians today have left behind this precious heritage of ours and move towards Western culture.

THIRTY-THREE
GANGAJAL

"If Indian culture is a coin then science and the traditions are two side of the coin."
-Akshay Bavda

"Gangajal" is considered to be the most sacred water in India. In India, almost everyone's house has a bottle filled with Ganga water. The river Ganga is also considered to be a very sacred river. In Indian culture it is believed that all the sins of the person who bathes in the river Ganga will be washed away, and heaven will be attained after death.

It is mentioned in ancient Indian culture that the river Ganga originates from the jata of Lord Shiva. Usually if the water is filled in a container it gets algae. As well as that water is no longer drinkable. But even if the Ganga water is stored for years, it does not spoil or grow algae. Hence Ganga water is considered holy and miraculous water.

Science is also behind such miraculous properties of Ganga water. The Ganga water has an abundance of viruses called "bacteriophages". The virus kills bacteria and algae, causing water pollution, and thus keeping the water pure for a long time. The bacteriophage is a virus that infects only bacteria and microbes, with no side effects on the human body. The Ganga has got some special properties, which are not found in other rivers. The special feature is that it also has self-cleaning features. Even today, after being so polluted, you can keep Ganga water in a bottle for decades and it does not rot or spoil.

Scientific experiments have shown that in most rivers, organic matter normally depletes the river's available oxygen. But in the Ganga, the discovery of an unknown substance or "X-Factor" that Indians call "disinfectant" has been proven, it processes organic matter and bacteria and kills them. The self-purifying quality of the Ganga emits 25 times more oxygen than other rivers in the world. Also, in some areas of the river Ganga, it has very high amount of sulphur, so bathing there can cures skin diseases.

Thus, the water of Ganga does not get polluted due to the reasons mentioned above. We can say that our sages were also adept in hydrology. They knew that flowing water had the property of self-purification so that in ancient Indian culture rivers were worshiped as mothers or goddesses.

Glossaries

- **Gangajal**– Water of Indian sacred river Ganga
- **Jata**– Tress/ Aigrette

About The Author

<u>AKSHAY BAVDA</u>

"**Akshay Bavda**" completed his M.Sc. from Bhavnagar University. As well as he awarded the Gold Medal for holding 1st rank in the University by Our Hon'ble Governor Shri Om Prakash Kohli sir. He is currently serving at Gujarat State Fertilizer and Chemical Limited. Apart from this, in the remaining time, he writes for his hobby.

www.ingramcontent.com/pod-product-compliance
Lightning Source LLC
Chambersburg PA
CBHW030913180526
45163CB00004B/1817